Humanoid Robots That Assemble Cars
XPENG'S AI REVOLUTION

A Deep Dive into the Ambitious Journey to Redefine Technology and Transform Urban Life

Tom K. Smith

Copyright ©Tom K. Smith, 2024.

All rights reserved. No part of this publication may be reproduced, distributed, or transmitted in any form or by any means, including photocopying, recording, or other electronic or mechanical methods, without the prior written permission of the publisher, except in the case of brief quotations embodied in critical reviews and certain other noncommercial uses permitted by copyright law.

Table of Contents

Introduction 3
Chapter 1: The Vision Behind Xpeng's AI Ambitions 7
Chapter 2: Introducing Iron – Xpeng's Humanoid Robot 15
Chapter 3: The Turing AI Chip – Xpeng's Technological Backbone 23
Chapter 4: The Kunlun Super Electric System and Energy Efficiency 34
Chapter 5: Xpeng's Autonomous Driving Platform – Kangai and Hawkeye 42
Chapter 6: Iron's Potential Beyond Manufacturing. 52
Chapter 7: Xpeng's Entry into Urban Air Mobility ... 63
Chapter 8: Xpeng's Global Expansion and Market Strategy 73
Chapter 9: China's Role in Xpeng's Rapid Growth .. 82
Chapter 10: The Future of AI-Driven Innovation at Xpeng 93
Conclusion 106

Introduction

Xpeng has emerged as one of the most intriguing players in the tech world, rapidly advancing from its early days in electric vehicle manufacturing to a full-fledged powerhouse of artificial intelligence and robotics innovation. Its journey has captivated industry watchers and consumers alike, as it takes bold steps that go far beyond traditional automaking. Initially recognized for its electric vehicles, Xpeng has set itself apart with a relentless drive to pioneer the integration of intelligent systems and groundbreaking technology in both familiar and uncharted spaces. What began as a venture focused on sustainable transportation has now grown into an expansive technological mission encompassing AI-powered humanoid robots, sophisticated AI chips, and even the prospect of personal air mobility through flying vehicles.

This book delves into Xpeng's ambitious and daring projects, each crafted to push the boundaries of what technology can do. From humanoid robots that support assembly lines and dream up new possibilities for customer service to custom-designed AI chips that power both vehicles and robotic entities, Xpeng's vision is bold. The company has stepped forward to redefine the expectations of what urban life, work, and travel might look like in the not-so-distant future. Each chapter will explore these projects in detail, providing a comprehensive look into how Xpeng's innovations fit within broader trends in AI, robotics, and mobility. Readers will discover not just the specifics of each advancement, but also the underlying strategy that binds these innovations together, creating a cohesive and visionary path forward.

Xpeng's role in the tech world is monumental, and its impact is felt on a global scale. In an industry dominated by tech giants from Silicon Valley to Shenzhen, Xpeng has carved a unique position by combining its mastery in electric vehicle production with a dedicated focus on artificial intelligence. By pushing forward with AI-based solutions that extend beyond the automotive sector, Xpeng challenges competitors, including well-established leaders like Tesla, and positions itself as a leader in an era where automation, smart machinery, and eco-friendly solutions are essential. This journey, marked by innovation and adaptability, highlights Xpeng's influence in reshaping industries and its commitment to changing the way we interact with technology in our daily lives.

As the world increasingly looks to AI for solutions to modern challenges, Xpeng stands out for its ability to transform complex, futuristic ideas into real, functioning products that meet tangible needs. This book will showcase Xpeng's dedication to making a meaningful difference, as seen through its expanding portfolio and its global aspirations. By the end, readers will gain a deeper understanding of how Xpeng's work touches on the fundamental shifts in technology and urban mobility, painting a picture of a future where AI and innovation are seamlessly woven into the fabric of everyday life.

Chapter 1: The Vision Behind Xpeng's AI Ambitions

Xpeng began with a straightforward yet ambitious goal: to make a mark in the electric vehicle industry by producing innovative, efficient, and user-centered cars that could compete with established global players. Founded with a focus on clean energy and advanced technology, Xpeng set out to create vehicles that would appeal to a new generation of eco-conscious consumers while incorporating a futuristic edge that was as much about intelligence as it was about mobility. From the outset, Xpeng's approach was distinctive; it wasn't just about producing another electric vehicle but about pushing the limits of what smart cars could offer. This vision quickly set Xpeng apart, emphasizing intelligent driving systems, immersive digital experiences, and a seamless integration of AI-powered features to deliver a unique customer experience.

As Xpeng grew within the EV sector, the company saw opportunities that extended beyond the automotive world. The advancements in artificial intelligence and robotics presented a chance to expand Xpeng's impact and deepen its technological expertise. Driven by a desire to redefine not only transportation but also the very fabric of daily life, Xpeng began exploring projects in robotics, AI chips, and even urban air mobility. The company understood that AI could play a transformative role, not only within vehicles but in diverse environments, from homes to workplaces and bustling city streets. Xpeng's leadership recognized that by combining automotive intelligence with robotics and AI applications, they could push the boundaries of how humans interact with technology on a broader scale.

This shift in focus marked a turning point for Xpeng, as it began to leverage its technical foundation in EV production to explore a broader realm of possibilities. The company invested in AI research and custom-built AI chips capable of supporting everything from autonomous driving systems to complex robotic functions. At the same time, Xpeng moved into robotics, developing humanoid robots with advanced dexterity, vision, and learning capabilities that went far beyond industrial applications. By expanding its focus to include these ambitious new fields, Xpeng began positioning itself as more than an automaker—it sought to become a leader in AI-driven technology, capable of influencing diverse industries and reshaping the urban landscape.

In essence, Xpeng's journey from a promising EV startup to a multifaceted technology company illustrates its forward-thinking vision and adaptability. By expanding into AI and robotics, Xpeng has broadened its scope and set itself on a path to influence sectors that go beyond transportation. This evolution underscores the company's commitment to innovation, demonstrating how its original goals have grown to encompass a future where intelligent machines seamlessly interact with and enhance human life.

As Xpeng established itself within the electric vehicle sector, its vision naturally expanded beyond just producing eco-friendly cars. Early success in EV manufacturing enabled the company to set more ambitious goals, recognizing the potential to revolutionize not only how people drive but how they interact with technology on a daily basis. This led Xpeng to shift from a purely automotive focus to a comprehensive technology strategy grounded in artificial intelligence. By integrating AI-driven solutions across multiple domains—humanoid robots, autonomous driving, and even personal air mobility—Xpeng set its sights on transforming urban life and reshaping the boundaries of mobility.

One of the first major shifts came with Xpeng's development of autonomous driving technology, a field already being pursued by other EV pioneers. However, Xpeng's approach was different. Rather than simply adding autonomous features to vehicles, Xpeng envisioned an entire ecosystem of AI-powered mobility. This vision included not only self-driving cars but also humanoid robots designed to assist in production and potentially other environments, as well as flying cars capable of tackling urban air travel. Xpeng's autonomous driving technology, driven by proprietary AI chips and advanced neural networks, was built to handle complex urban environments, pushing toward a future where people could rely on fully autonomous vehicles in both personal and commercial contexts. The commitment to AI integration soon became a core component of Xpeng's strategy, shaping its approach to technology in a way that went far beyond traditional automotive goals.

In the global tech landscape, Xpeng's multi-pronged approach is both bold and distinctive. Comparisons are often drawn with industry leaders like Tesla, which also has its sights set on autonomous driving and AI. Yet, while Tesla focuses on applying its autonomous technology directly to consumer vehicles, Xpeng has pursued a more expansive vision that encompasses different types of AI-powered innovations. Tesla's Optimus robot project, for instance, aims to produce humanoid robots that could support people in tasks beyond driving, while Xpeng's own humanoid robot, Iron, operates with similar intentions yet extends its applications directly into manufacturing and service-oriented roles. Xpeng's broader focus on robotics, custom AI chips, and even personal flying vehicles reflects a commitment to creating an interconnected technology ecosystem that not only powers cars but also touches upon multiple aspects of urban living.

By pushing the boundaries of mobility to include flying vehicles, Xpeng sets itself apart in ways few competitors have ventured. Its development of EVTOL (electric vertical takeoff and landing) vehicles aims to introduce urban air mobility solutions, something that remains largely unexplored by Tesla and other EV companies.

Xpeng's flying car prototypes, equipped with sophisticated AI systems for navigation and safety, represent an effort to create new dimensions of transport that bypass conventional roads and directly address urban congestion challenges. This expansion into air travel demonstrates Xpeng's forward-thinking vision, leveraging AI not only to improve ground transportation but to make the dream of flying cars a realistic option within urban settings.

In essence, Xpeng's evolution from EV manufacturing to a full-scale AI tech company marks a bold and distinctive strategy in the tech world. By exploring fields beyond traditional automotive innovation, Xpeng is positioning itself as a pioneer of the next frontier in AI-driven technology. While Tesla and others focus heavily on perfecting autonomous ground transportation, Xpeng envisions a future where AI-driven robots, flying vehicles, and self-navigating systems work together to create a seamless, interconnected urban experience. This unique approach has placed Xpeng in an influential position among global tech leaders, illustrating how it stands apart with a vision that goes beyond cars, reaching into the future of how technology integrates into every aspect of life.

Chapter 2: Introducing Iron – Xpeng's Humanoid Robot

Xpeng's humanoid robot, Iron, is a marvel of design and engineering, crafted with meticulous attention to both form and functionality. Standing at 5 feet 8 inches (or 178 centimeters) and weighing approximately 154 pounds (70 kilograms), Iron closely mirrors the dimensions of an average adult human. This deliberate choice in size makes Iron agile enough to navigate diverse environments while remaining sturdy and capable of performing a wide range of tasks, from heavy lifting to precise assembly work. Xpeng designed Iron to serve not just as a powerful machine but as a flexible assistant capable of mimicking human motions with remarkable precision.

A key feature of Iron's design is its complex joint structure, which allows for a high degree of articulation and freedom of movement. The robot boasts over 60 joints, enabling it to replicate human-like motions with ease. These joints grant

Iron a total of 200 degrees of freedom, a remarkable level of agility that allows it to adapt fluidly to various tasks in a factory setting or other environments. Each joint is carefully calibrated to perform actions that are both swift and controlled, ensuring that Iron can handle delicate parts without damaging them, while also being able to support heavier tasks that require strength. This range of movement transforms Iron into a versatile tool that can perform tasks once thought exclusive to human workers.

Iron's hands, in particular, showcase an impressive degree of precision, featuring 15 degrees of freedom in each hand alone. This advanced hand functionality enables Iron to grip, lift, position, and place objects with a finesse similar to that of a human. From grasping small components to handling larger parts, Iron's hands are capable of adjusting their grip to the task at hand, making the robot suitable for intricate assembly work where accuracy is paramount. This precision in hand design allows Iron to manage multiple tools and objects in real time, giving it the ability to seamlessly transition between different assembly functions within a manufacturing process.

The overall movement capacity of Iron is enhanced by Xpeng's proprietary Turing AI chip, which powers the robot's decision-making and adaptive capabilities. This chip processes Iron's environment, enabling it to respond in real-time to changes and make adjustments as needed. This capacity for environmental awareness allows Iron to walk, pivot, bend, and extend its limbs with precision and confidence, functioning effectively on the factory floor. Whether it's walking around obstacles, repositioning itself in a limited space, or picking up specific components in a dynamic setting, Iron's design is built to handle the complexities of a constantly evolving workspace.

Together, these physical attributes make Iron a sophisticated piece of technology that goes beyond traditional robotics. The combination of human-like proportions, finely tuned joints, and a high degree of movement precision reflects Xpeng's vision of a robot capable of integrating into human environments with ease. This thoughtful design enables Iron to function not only as an assembly assistant in production lines but also as a potential service robot in settings like retail or customer assistance, showing how its robust yet flexible design can adapt to a variety of roles beyond manufacturing.

Iron's role in Xpeng's production line marks a significant step toward the future of automated manufacturing, where robots and humans can work seamlessly together. At the heart of Xpeng's assembly line, Iron is tasked with assembling the P7 Plus electric vehicle, a process that requires both precision and adaptability. Xpeng's vision for Iron goes beyond simple automation; the robot is designed to perform tasks that typically require human dexterity and attention to detail. Integrated into various stages of the assembly, Iron can be seen handling intricate parts, securing components, and adjusting its approach based on the specific requirements of each step. This ability to adapt its movements with human-like precision makes Iron an invaluable asset, particularly for high-quality assembly work where consistency and accuracy are essential.

Iron's integration into the production line is made possible by the advanced processing power of Xpeng's custom-built Turing AI chip. This chip is the core of Iron's decision-making and control systems, effectively acting as the robot's "brain." With a powerful 40-core processor designed specifically for AI-driven applications, the Turing AI chip equips Iron with the ability to process vast amounts of data in real time. It supports complex AI models with up to 30 billion parameters, giving Iron the flexibility to interpret sensory inputs, analyze its surroundings, and adapt to changing conditions on the factory floor. This capacity allows Iron to move seamlessly between tasks, manage multiple variables simultaneously, and handle unexpected adjustments with ease.

The Turing AI chip was created not only for robotics but also for Xpeng's entire suite of AI-powered technologies, including autonomous driving systems. By integrating this chip into Iron, Xpeng has enhanced the robot's ability to reason, plan, and adapt in a way that goes far beyond traditional robotic programming. Rather than following a fixed set of instructions, Iron can process its environment dynamically, allowing it to make decisions based on real-time data. This adaptability is crucial in an industrial setting where precision and responsiveness are required, as Iron must be able to respond immediately to any changes or new tasks assigned by the production line.

Thanks to the Turing AI chip, Iron operates with an impressive level of autonomy and intelligence. The chip enables Iron to execute complex maneuvers, like adjusting grip pressure, maintaining balance, and coordinating its limbs in tight spaces. This makes it an ideal assistant for assembling delicate components or handling heavy parts with care. For example, when assembling the P7 Plus, Iron can switch effortlessly between tasks, whether it's positioning a bolt, aligning panels, or applying specific pressure to ensure components are securely fixed. The chip's processing power allows Iron to "learn" from repetitive actions, refining its movements and improving efficiency over time.

The Turing AI chip's role in powering Iron reflects Xpeng's broader vision of integrating advanced AI into its products, from autonomous vehicles to humanoid robots. This custom-built chip is designed to meet the demanding needs of AI applications, providing the computational resources that enable Iron to function as a truly adaptive and intelligent machine. By equipping Iron with this powerful chip, Xpeng has created a robot that can operate autonomously in complex settings, bridging the gap between AI research and real-world application. Iron's role in the production line is just the beginning; as Xpeng continues to refine the Turing AI chip and expand its capabilities, Iron may evolve to take on even more diverse roles across industries, potentially transforming the nature of robotic assistance in both manufacturing and beyond.

Chapter 3: The Turing AI Chip – Xpeng's Technological Backbone

The Turing AI chip that powers Iron is a cutting-edge piece of technology, meticulously engineered to support the rigorous demands of artificial intelligence and machine learning applications in both robotics and autonomous vehicles. At the core of its performance is a powerful 40-core processor, designed to handle complex computational tasks with remarkable speed and efficiency. Each core operates in tandem to maximize processing power, allowing the chip to handle multiple high-intensity operations simultaneously. This multi-core architecture is crucial for ensuring that Iron can perform real-time data processing, environmental analysis, and adaptive responses without delay, even in fast-paced, dynamic production settings.

One of the Turing chip's standout features is its capacity to manage up to 30 billion parameters, a level of complexity that pushes it into the realm of advanced AI models. Handling such a large number of parameters means that the chip can support machine learning models capable of nuanced decision-making and learning from experience. For Iron, this parameter capacity translates to a more sophisticated understanding of its environment and the ability to adapt its movements and responses based on real-time feedback. This allows Iron to navigate complex assembly tasks, adjust its grip on various materials, and operate with human-like precision in settings where even slight adjustments can impact the quality and accuracy of its work.

Beyond its processing power and parameter-handling capabilities, the Turing chip also includes specialized neural network accelerators. These accelerators are designed to optimize the processing of deep learning algorithms, significantly enhancing Iron's ability to interpret sensory data and make complex calculations at lightning speed. By using dedicated hardware to run deep learning tasks, the Turing chip reduces computational load on the main processor, ensuring that Iron can execute rapid decision-making processes without lag. This feature is especially valuable in the manufacturing environment, where Iron must continually assess and respond to its surroundings, recognize objects, and coordinate its actions based on constantly updated information.

The Turing chip's architecture also includes advanced memory management, designed to handle vast amounts of data seamlessly. This high memory bandwidth allows Iron to retrieve and process large datasets in real time, enabling it to react swiftly to changes and maintain high performance even under heavy workloads. In an environment like a production line, where the ability to handle repetitive tasks without error is critical, this efficient memory handling ensures that Iron can keep pace with the demands of modern manufacturing. It also supports complex decision trees, allowing the robot to autonomously perform sequences of actions and maintain high levels of productivity.

Another unique feature of the Turing AI chip is its ability to operate with low latency, a critical factor in time-sensitive applications. Low latency means that Iron can process data and execute actions almost instantaneously, an essential quality in a robotic assistant that needs to react quickly to changes in its environment. In practice, this enables Iron to respond to last-minute adjustments on the assembly line, recognize and correct minor alignment issues, and handle variations in parts with fluidity and precision. The chip's low latency further supports Iron's potential for applications beyond the factory floor, making it adaptable to roles where quick, adaptive responses are crucial, such as customer service or healthcare assistance.

The Turing AI chip represents the culmination of Xpeng's investment in creating an AI processor that can support a wide range of applications, from robotics to autonomous driving. By integrating such advanced specifications, including its 40-core processor, 30-billion-parameter capacity, neural network accelerators, high memory bandwidth, and low-latency operation, Xpeng has crafted a powerhouse chip that equips Iron with the intelligence and adaptability necessary for sophisticated tasks. This technology positions Iron not only as a vital asset on the assembly line but as a potential trailblazer in the broader field of AI-driven robotics. As Xpeng continues to develop the Turing chip's capabilities, it holds the potential to shape the future of AI applications across multiple industries, expanding the possibilities of what robots like Iron can achieve in both industrial and everyday environments.

The Turing AI chip serves as the backbone of Xpeng's expanding array of intelligent technologies, uniting robotics, autonomous vehicles, and the future vision of flying cars under a single, powerful AI system. Designed with flexibility at its core, the Turing chip seamlessly transitions between supporting robots like Iron on the production line and powering autonomous vehicle functions. This adaptability is key to Xpeng's strategy of creating a cohesive technological ecosystem where each AI application shares a foundational intelligence, allowing for efficient communication and coordination across platforms. By supporting robotics, ground-based autonomous vehicles, and future air mobility, the Turing chip exemplifies Xpeng's ambition to build an integrated network of AI-driven technologies capable of transforming urban life and transportation.

In robotics, the Turing chip enhances Iron's ability to perform diverse, adaptive tasks on the assembly line. The chip's processing power allows Iron to not only replicate human-like movements but also to learn and refine these movements over time, using data gathered from its interactions to improve precision and efficiency. Beyond performing repetitive tasks, Iron's decision-making capabilities are elevated by the Turing chip, which enables it to make real-time adjustments, detect variances in materials, and respond intuitively to minor changes on the factory floor. This dynamic responsiveness is essential in a production environment where every adjustment and refinement can impact the quality and speed of manufacturing. The chip's high capacity for handling parameters also means Iron can handle a vast number of instructions simultaneously, maintaining efficiency even in complex sequences of actions.

In Xpeng's autonomous vehicles, the Turing chip plays an equally crucial role, enabling them to achieve a high level of autonomy. The chip powers Xpeng's proprietary neural network for autonomous driving, designed to process the immense amounts of sensory data that a vehicle must interpret to operate safely without human intervention. By leveraging deep learning algorithms optimized by the Turing chip's neural network accelerators, Xpeng's autonomous vehicles can recognize and respond to complex traffic scenarios, obstacles, and environmental conditions with impressive accuracy. This ability to interpret and act upon data in real time enables Xpeng's vehicles to adapt dynamically to changing road conditions, making them safer and more reliable in diverse driving environments. The chip's processing power further enhances these vehicles' situational awareness, giving them the ability to track objects, detect pedestrian movement, and predict the behavior of other vehicles, creating a holistic and responsive autonomous driving experience.

Xpeng's vision for the Turing chip doesn't stop on the ground; it extends into the realm of urban air mobility with the development of flying vehicles. By applying the chip's robust AI capabilities to EVTOL (electric vertical takeoff and landing) vehicles, Xpeng envisions a future where personal air travel becomes part of everyday mobility. The Turing chip's low-latency response and high processing power make it an ideal choice for controlling flying vehicles, where split-second decisions are critical for safety in the air. For EVTOLs, the chip would enable autonomous navigation, route optimization, and precise control of the vehicle's flight dynamics, all while managing the complex data streams required to maintain stable, safe operations in the sky. With the Turing chip's support, Xpeng's flying cars could autonomously navigate urban airspace, creating new possibilities for congestion-free transportation in dense metropolitan areas.

The Turing AI chip also advances the autonomy and decision-making capabilities of Xpeng's technologies across all applications, allowing Iron and autonomous vehicles alike to learn, adapt, and perform complex tasks with remarkable efficiency. This capacity for learning is built into the chip's architecture, enabling both robots and vehicles to analyze their actions, optimize for greater efficiency, and even "learn" from experience. In Iron's case, this means refining its assembly techniques, learning the optimal grip for specific components, or adjusting its movements to avoid potential errors. For Xpeng's autonomous vehicles, this adaptability means analyzing past driving data to improve future performance, handling unexpected road conditions, and responding to new driving scenarios without requiring additional programming.

By fostering autonomy across Xpeng's products, the Turing chip sets the stage for an intelligent, self-improving network of machines that grow smarter with every interaction. The ability to continuously process and analyze data also positions these technologies to make increasingly sophisticated decisions, creating a foundation for true AI-driven mobility. Xpeng's commitment to advancing autonomy through the Turing chip demonstrates a vision of technology that is capable not just of performing tasks but of evolving over time. With its adaptability across diverse AI applications, the Turing AI chip is a critical component of Xpeng's future, supporting a range of autonomous functions that promise to redefine how we live, work, and travel.

Chapter 4: The Kunlun Super Electric System and Energy Efficiency

The Kunlun system is a cornerstone of Xpeng's electric vehicle technology, designed to enhance the efficiency, range, and overall performance of its EVs. At the heart of the Kunlun system is high-voltage silicon carbide (SiC) technology, an advanced material known for its exceptional durability, heat resistance, and electrical conductivity. By leveraging SiC technology, Xpeng has been able to significantly optimize energy conversion within its vehicles, minimizing power loss and boosting the overall efficiency of the powertrain. This advancement plays a pivotal role in extending the range of Xpeng's EVs, addressing one of the most critical concerns for electric vehicle owners: the need to maximize distance on a single charge.

Silicon carbide is a game-changer in the world of EVs, as it allows the Kunlun system to operate at higher voltages and temperatures compared to traditional silicon-based systems. This enables more effective power management, especially during demanding situations like acceleration or long-distance driving, where power efficiency directly impacts performance and range. By incorporating SiC technology into the Kunlun system, Xpeng has made its EVs more capable of meeting the high expectations of modern drivers who demand both sustainability and convenience. This high-voltage capability also allows for faster and more efficient charging, reducing the time a vehicle needs to reach a full or near-full charge, which is another significant factor for drivers concerned with minimizing downtime.

The Kunlun system's ability to support high-voltage operations not only boosts the vehicle's range but also contributes to a smoother and more responsive driving experience. This system manages the EV's power delivery with precision, ensuring that the vehicle can accelerate smoothly and handle various driving conditions with ease. As a result, drivers experience better performance without sacrificing energy efficiency. The system's careful control over power distribution within the vehicle helps maintain battery health over time, making the EV a more sustainable and long-lasting choice. By harnessing the full potential of silicon carbide, the Kunlun system positions Xpeng's EVs at the forefront of energy-efficient technology, helping the company stand out in an increasingly competitive market.

The role of the Kunlun system extends beyond just energy efficiency; it also impacts the practicality of EV ownership by reducing charging time significantly. With Kunlun's integration, Xpeng's EVs can charge from 10% to 80% in just 12 minutes when connected to a compatible high-speed charging station. This breakthrough in charging speed addresses a key concern among electric vehicle users: the need to balance convenience with eco-friendly transportation. For drivers, the reduced charging time means less waiting and more time on the road, making EVs a more viable option for longer trips and reducing the range anxiety often associated with electric driving. This fast-charging capability, combined with extended range, positions Xpeng's EVs as convenient, reliable options for drivers transitioning from traditional combustion engines to electric power.

The Kunlun system is not only a technical enhancement but a strategic advancement for Xpeng as it competes in the global EV market. By implementing silicon carbide technology, Xpeng demonstrates a commitment to innovation that extends beyond the basic requirements of electric mobility. Kunlun embodies Xpeng's mission to build a comprehensive EV experience, one that balances performance, sustainability, and user satisfaction. In doing so, the system plays a central role in helping Xpeng establish itself as a leader in the electric vehicle industry, offering drivers not just a greener alternative but a technologically advanced driving experience that rivals traditional vehicles. As Xpeng continues to develop and refine its EV technology, the Kunlun system will likely serve as a foundation for further advancements, paving the way for more efficient, powerful, and user-friendly electric vehicles in the future.

The Kunlun system represents a breakthrough in electric vehicle performance, delivering significant enhancements to both driving range and charging efficiency. Through its integration of high-voltage silicon carbide technology, Kunlun has enabled Xpeng's EVs to reach an impressive range of up to 870 miles on a single charge. This extended range offers a transformative shift for electric vehicle users, providing the freedom to travel farther without the constant need to recharge. For many drivers, this range significantly exceeds the capabilities of other EVs on the market, allowing for a greater sense of reliability and convenience, especially on longer journeys.

The Kunlun system's high-voltage design not only extends range but also powers an ultra-fast charging capability. Xpeng's EVs equipped with this system can go from 10% to 80% charge in a mere 12 minutes when using compatible high-speed chargers. This leap in charging efficiency means that EV drivers can now achieve a substantial recharge during a short break, making long trips less of a logistical challenge. By reducing the downtime associated with charging, Kunlun makes EV ownership more compatible with the needs of busy drivers who depend on quick stops to maximize time on the road. This innovation also appeals to users in urban areas, where access to rapid charging options can streamline daily commuting and travel routines.

For long-distance travel, these enhancements are game-changing. Range anxiety—the fear that an EV will run out of power before reaching a charging station—has been one of the main obstacles deterring some drivers from switching to electric vehicles. With Kunlun's extended range, EV drivers have the assurance that they can complete long trips with minimal recharging interruptions, helping to build confidence in electric mobility as a reliable option for all kinds of journeys. This breakthrough is particularly valuable in regions where charging infrastructure is still developing; with Kunlun-equipped EVs, drivers can comfortably manage longer routes, knowing that fewer stops will be required. The reduced need for frequent charging also decreases the dependency on a dense network of charging stations, broadening the appeal of EVs to a wider range of users, including those in more remote or less urbanized areas.

The Kunlun system's fast-charging capability further enhances the practicality of EVs for long-distance travelers. With a quick recharge possible in just 12 minutes, drivers can maintain momentum during extended road trips without lengthy waits. This convenience mirrors the quick refueling experience of traditional combustion vehicles, helping to bridge the gap for users accustomed to gasoline-powered cars. By offering both an extended range and rapid charging, Xpeng addresses two of the most pressing concerns among prospective EV buyers, making its vehicles an attractive choice for those who value performance and convenience.

In essence, the range enhancements and fast-charging capabilities of the Kunlun system make EVs more viable for a broader audience. Long-distance travelers, commuters, and drivers in regions with limited charging access can all benefit from these advancements, which transform the electric vehicle experience from one limited by infrastructure and range into one marked by freedom and flexibility. With Kunlun, Xpeng has not only improved the technological foundation of its EVs but has also positioned itself as a leader in the push to make electric vehicles a practical choice for all types of drivers, paving the way for a future where range anxiety and charging delays are obstacles of the past.

Chapter 5: Xpeng's Autonomous Driving Platform – Kangai and Hawkeye

The Kangai platform represents a significant advancement in Xpeng's pursuit of full autonomy, bringing the company's vehicles to the brink of Level 4 autonomous driving. This level of autonomy is characterized by the vehicle's ability to perform all driving tasks without human intervention in most situations, marking a leap beyond the Level 2 and Level 3 systems that still require some driver oversight. With Kangai, Xpeng has developed a system that can handle the complexities of real-world driving, effectively transforming the vehicle into a self-sufficient navigator. This shift holds immense potential, especially for the deployment of robotaxis, where human-free driving is essential for maximizing efficiency, safety, and accessibility in urban environments.

Kangai's Level 4 autonomy capabilities are powered by a sophisticated network of sensors, cameras, radar, and LIDAR, all of which work together to give the vehicle a complete, 360-degree awareness of its surroundings. These sensory inputs feed into a high-powered processing unit, enabling Kangai to analyze and react to various driving conditions in real time. The system interprets data on everything from road signs, lane markings, and pedestrian crossings to complex traffic scenarios, including high-speed highway driving and congested urban streets. This ability to "see" and understand the environment allows the vehicle to make informed decisions without relying on human input, effectively taking control in a way that prioritizes safety and accuracy.

Kangai's advanced AI algorithms enable it to predict and adapt to the behaviors of other drivers, pedestrians, and cyclists, creating a driving experience that is not only autonomous but also intuitively responsive to the dynamics of traffic. For instance, Kangai can anticipate when a pedestrian might cross the street or when another vehicle may switch lanes abruptly, allowing it to adjust speed or lane position to avoid potential conflicts. This predictive capacity is crucial for achieving the reliability required of Level 4 autonomy, where the vehicle must be capable of handling unexpected scenarios autonomously. In addition, Kangai continuously learns from driving data, refining its responses over time and improving its capacity to navigate increasingly complex environments.

The capabilities of Kangai have significant implications for Xpeng's robotaxi initiative. Robotaxis equipped with Kangai's Level 4 autonomy can operate without a human driver, making them highly efficient for urban transportation networks. These autonomous taxis are designed to transport passengers seamlessly across cities, offering a convenient, on-demand mobility solution that reduces reliance on personal vehicles and contributes to easing urban congestion. Without the need for a driver, these robotaxis can operate continuously, providing flexible transport options day and night, with minimal human intervention. Kangai's technology enables the vehicle to recognize and adapt to specific routes, pick-up and drop-off locations, and traffic regulations, offering a consistent and safe service for users.

The deployment of Kangai-powered robotaxis could revolutionize public and private transportation, especially in densely populated areas where parking and traffic congestion are constant challenges. By allowing for autonomous, shared transportation, Kangai presents a sustainable mobility option that could reduce the environmental footprint of urban commuting and ease the burden on road infrastructure. Additionally, the increased safety provided by Level 4 autonomy—where human error is virtually eliminated—positions Kangai's robotaxis as a potentially safer alternative to traditional ride-hailing services.

With Kangai, Xpeng is not only advancing the technology behind autonomous driving but also shaping the future of urban mobility. Level 4 autonomy signifies a shift towards a world where cars can operate independently, transforming transportation into a service that is both accessible and efficient. Kangai's capabilities make this vision a tangible reality, offering a glimpse into a future where robotaxis equipped with advanced autonomy redefine how people move through cities. By bridging the gap between human-driven and fully autonomous vehicles, Kangai sets a new standard in the evolution of intelligent transportation, highlighting Xpeng's commitment to innovation in the AI-driven automotive space.

The Hawkeye Pure Vision system is a revolutionary component in Xpeng's autonomous vehicle technology, designed to provide unparalleled awareness and precision. By equipping vehicles with a 720-degree field of vision, the Hawkeye system enables Xpeng's robotaxis to "see" their surroundings with extraordinary clarity, eliminating blind spots that are common in conventional vehicles. This expansive view is achieved through a sophisticated array of sensors, cameras, radar, and LIDAR, strategically positioned around the vehicle to capture a comprehensive picture of the environment. This complete visual coverage enables the vehicle to make informed decisions, navigate complex traffic situations, and avoid potential obstacles from every angle.

With Hawkeye's 720-degree vision, Xpeng's robotaxis can detect vehicles, pedestrians, cyclists, and other obstacles across all lanes of traffic, including those that may approach from behind or the sides. This capability is crucial in high-density, urban environments where unpredictability is a constant factor, and having complete spatial awareness helps the vehicle anticipate and respond to any sudden changes. For instance, if a cyclist suddenly veers into the robotaxi's path from the rear or if a pedestrian steps into the road from a blind corner, the Hawkeye system's wide field of vision allows the vehicle to recognize these movements instantly and react appropriately. By eliminating blind spots, the Hawkeye Pure Vision system greatly enhances the robotaxi's ability to maneuver safely in all types of conditions, making it an essential component for achieving true autonomy.

The enhanced safety features of Xpeng's autonomous vehicles go hand-in-hand with the capabilities of the Hawkeye system. Integrated safety mechanisms are designed to protect passengers in all scenarios, whether during routine navigation or unexpected situations. For example, the robotaxi can perform controlled emergency braking if it detects an imminent collision, instantly slowing down or coming to a stop if a pedestrian or vehicle suddenly enters its path. This reaction is possible because the Hawkeye system continuously monitors and interprets its environment, feeding real-time data to the vehicle's AI, which then makes rapid calculations to determine the best course of action.

To further enhance safety, Xpeng's autonomous vehicles are equipped with predictive algorithms that analyze and anticipate the behaviors of nearby vehicles and pedestrians. These algorithms enable the robotaxi to predict if a neighboring car might make an unexpected lane change or if a pedestrian might cross the street without warning. By analyzing these behaviors in advance, the vehicle can adjust its speed, position, and trajectory to maintain a safe distance, actively avoiding situations that could lead to accidents. This proactive approach to safety marks a significant improvement over traditional driving, where a driver's field of vision and reaction times are often limited. With Hawkeye's extensive visual range and predictive algorithms, the robotaxi continuously adapts to its surroundings, creating a driving experience that is both safer and more reliable.

Another key feature of Xpeng's safety framework is redundancy in critical systems. This design ensures that if one sensory or processing component fails, a backup system takes over immediately, allowing the vehicle to continue operating safely until the issue can be resolved. This redundancy is essential in driverless scenarios, where human intervention is not an option. By maintaining a multi-layered safety system, Xpeng minimizes the risk of system failure impacting passenger safety. Coupled with the 720-degree coverage provided by Hawkeye, these redundancy measures create a robust safety net that reassures passengers and instills confidence in autonomous transportation.

In driverless scenarios, Xpeng's autonomous vehicles are also programmed to respond to external safety signals, such as emergency vehicles or unexpected obstacles. The Hawkeye system can identify flashing lights or sirens from emergency vehicles, prompting the robotaxi to yield or pull over as necessary. This situational awareness helps the vehicle comply with traffic laws and ensures that it operates responsibly, even in scenarios where quick adjustments are required. Additionally, the vehicle's software is regularly updated to account for new safety protocols and local regulations, ensuring that it remains compliant and capable of handling the latest urban safety challenges.

Overall, the Hawkeye Pure Vision system and Xpeng's integrated safety features work together to create a comprehensive safety strategy that prioritizes passenger protection and reliable autonomous driving. With 720-degree vision, enhanced predictive capabilities, and built-in redundancies, Xpeng's robotaxis are engineered to offer a secure, seamless experience for passengers in every situation. This safety-focused approach not only advances Xpeng's vision of autonomous urban mobility but also sets a new standard in autonomous vehicle safety, paving the way for a future where driverless technology is trusted and embraced by the public.

Chapter 6: Iron's Potential Beyond Manufacturing

Xpeng's humanoid robot, Iron, has the potential to become a transformative force in customer service and retail environments. Built with human-like proportions, dexterity, and advanced AI capabilities, Iron is uniquely suited for tasks that require interaction, adaptability, and precision. While initially designed for manufacturing and assembly on the production line, Iron's design and intelligence position it for broader applications in settings where customer engagement, support, and assistance are essential.

In retail spaces, Iron could become a seamless part of the customer experience, providing hands-on assistance that goes beyond traditional customer service roles. Imagine walking into a store where Iron greets customers, answers questions, and helps them find specific products. With its advanced sensor system, Iron can recognize individual customers and respond to their inquiries with tailored information, such as the location of specific items or personalized product recommendations based on past interactions. Iron's human-like mobility and dexterity allow it to perform tasks that involve physical engagement, such as retrieving products from shelves, organizing displays, or even restocking items in real time. This adaptability makes Iron a valuable asset for busy retail environments, where staff availability can fluctuate, and efficient customer service is crucial.

Iron's functionality in retail can extend to interactive product demonstrations. For example, Iron could show customers how to use a new device, guiding them through its features and functionalities in a hands-on demonstration. By using its articulate limbs and precise hand movements, Iron could handle delicate items, demonstrating products in a way that both informs and engages customers. This type of interactive service adds a level of engagement and depth that is difficult to achieve with static displays or digital kiosks, making the shopping experience more immersive and memorable. Iron's AI could also learn customer preferences over time, allowing it to provide increasingly personalized assistance based on past interactions, contributing to brand loyalty and enhancing the overall shopping experience.

Beyond retail, Iron has significant potential in office environments, where its versatility can support a range of administrative and operational tasks. In an office setting, Iron could function as a high-tech assistant, greeting visitors at the reception area, guiding them to meeting rooms, and even managing appointment schedules. Iron's natural language processing abilities would allow it to engage in simple, conversational interactions, making it a welcoming presence for guests and clients. Furthermore, Iron's ability to recognize and interpret human cues would enable it to adapt its approach based on the visitor's needs, offering a level of attentiveness that enhances the professionalism of the workplace.

Iron could also be programmed to assist with repetitive office tasks, such as organizing files, delivering documents, or setting up meeting spaces. With its ability to learn from routines and optimize its actions, Iron can manage tasks efficiently, freeing up human employees to focus on more complex work. In dynamic office environments, where multitasking and quick adaptation are often required, Iron's ability to adjust to various tasks and work alongside people would make it a reliable partner in maintaining office productivity.

In the long term, Iron's presence in retail and office settings could reshape how we think about customer service and workplace assistance. The robot's ability to perform both physical and cognitive tasks with precision and adaptability allows it to bridge the gap between automated services and personal interaction. Unlike traditional automated systems, Iron's human-like appearance and interactive skills foster a more relatable and engaging experience for customers and clients alike.

By blending technological sophistication with a user-centered approach, Iron exemplifies how advanced robotics can be integrated into everyday life, enhancing service standards and elevating customer experiences across diverse environments.

As businesses increasingly look to technology for operational support, Iron's application in retail and customer service signals a future where intelligent robots are not only tools but also part of the human experience. With its adaptability and capacity to learn, Iron could contribute to a world where high-quality service and support are accessible, responsive, and consistent, paving the way for a new era in customer engagement and workplace efficiency.

AI-powered robots like Xpeng's Iron are set to redefine everyday life by enhancing customer service, streamlining task management, and offering highly personalized assistance. As humanoid robots become more capable and adaptable, they present opportunities to transform a wide array of industries, from retail to healthcare, hospitality, and beyond. In customer-facing roles, Iron can assist by providing tailored recommendations, offering product information, or guiding customers through a store, creating a service experience that is both efficient and engaging. Unlike traditional automated solutions, Iron can interact naturally, recognizing customer preferences, learning from each interaction, and adjusting its approach to deliver truly personalized assistance.

In workplaces, Iron's role expands to managing essential tasks that support daily operations. In office environments, Iron could assist with duties like greeting visitors, directing them to appointments, and performing light administrative tasks such as organizing files or delivering messages. This allows human employees to focus on more strategic responsibilities while ensuring that basic, yet essential, tasks are handled with consistency and reliability. For industries requiring repetitive but high-precision tasks, such as logistics or healthcare, Iron could be a dependable assistant, handling everything from inventory management to patient assistance with a high level of accuracy. These applications show that Xpeng's humanoid robots have the potential to bridge the gap between human-driven tasks and traditional automation, creating a versatile workforce that enhances both efficiency and user experience.

In the competitive landscape of humanoid robotics, Xpeng's Iron shares certain similarities with other major robots, such as Tesla's Optimus. Both Iron and Optimus are designed as humanoid robots intended to take on versatile roles in manufacturing and customer service. They are equipped with advanced AI and processing capabilities that enable them to move with agility, perform delicate tasks, and respond dynamically to changes in their environment. Like Optimus, Iron can perform repetitive tasks with high precision, making it suitable for manufacturing environments where consistency and efficiency are key. Both robots are designed to mimic human-like gestures, allowing them to interact with people in a relatable manner, which is especially valuable in customer service and retail settings.

However, there are also key differences in the design philosophy and functionality of Iron compared to its competitors. Tesla's Optimus, for instance, draws heavily on Tesla's autonomous driving technology and is designed to operate in a wide range of environments, with an emphasis on affordability and mass production. Optimus aims to be a general-purpose robot that can be applied to tasks across various sectors, from home use to heavy industry. Xpeng's Iron, on the other hand, is initially specialized for production line assistance in Xpeng's own automotive manufacturing and is equipped with a proprietary Turing AI chip that tailors its capabilities for specific applications within the Xpeng ecosystem. This focused application allows Iron to excel in environments requiring dexterity and precision, particularly in settings where it works alongside humans.

In comparison to other humanoid robots emerging in the market, such as those developed by companies like Boston Dynamics, Iron represents a middle ground between specialized functionality and versatility. Boston Dynamics' robots are often designed with an emphasis on physical agility and advanced mobility, making them suitable for complex terrains and tasks requiring robust movement. Iron, while less focused on acrobatic mobility, is highly proficient in handling objects, interacting with people, and performing specific tasks that require both accuracy and adaptive learning. This balance between dexterity and functionality positions Iron as a practical solution for roles that require fine motor skills and adaptability rather than extreme physical capabilities.

The development of Iron and other humanoid robots highlights an evolving competition in the AI and robotics space, where companies are defining the roles these robots will play in everyday life. Xpeng's approach with Iron reflects a vision of creating AI-powered assistants that can integrate into human-centric spaces, helping customers, employees, and consumers alike. By focusing on customer service, task management, and personalization, Xpeng's Iron stands out for its potential to interact with people in meaningful, helpful ways, bridging the gap between human services and automated solutions. In contrast, Tesla's Optimus and other competitors are designed with a broader focus, aiming to function as all-purpose robots adaptable to a wider range of tasks and industries.

Ultimately, Xpeng's focus on specific applications and the integration of a custom AI system with Iron enables it to serve as an efficient, specialized tool in areas that are high-touch and service-oriented. As humanoid robots continue to develop, Iron's unique position highlights the varied directions that

AI-powered robotics can take, each bringing a different vision of how robots might shape the future of work, customer interaction, and day-to-day assistance. With technology advancing rapidly, the competitive landscape in humanoid robotics will likely see these robots evolving in ways that make them increasingly accessible, responsive, and essential to both businesses and everyday users.

Chapter 7: Xpeng's Entry into Urban Air Mobility

Xpeng's AEROHT division represents the company's ambitious step into the realm of urban air mobility, pioneering electric vertical takeoff and landing (EVTOL) vehicles that could transform how people navigate cityscapes and dense urban areas. AEROHT, as a subsidiary dedicated to advancing the next generation of transportation, focuses on developing EVTOLs that seamlessly integrate with Xpeng's vision of sustainable, AI-driven mobility. With EVTOL technology, AEROHT aims to offer a viable solution for urban congestion, providing an alternative that doesn't rely on traditional road infrastructure and that could redefine what it means to commute in modern cities.

AEROHT's EVTOLs are designed to be electric-powered and equipped with vertical takeoff and landing capabilities, allowing them to operate in tight spaces where traditional runways aren't feasible. This makes them particularly suited for urban environments, where compact design and minimal environmental footprint are essential. By utilizing electric power, AEROHT's EVTOLs align with Xpeng's commitment to sustainability and reduced emissions, offering a cleaner alternative to conventional aircraft and even ground-based vehicles. The transition to electric propulsion not only reduces noise pollution but also contributes to a more eco-friendly urban transit solution, which could benefit densely populated areas struggling with high levels of vehicle emissions.

Xpeng's AEROHT team brings together experts in aerodynamics, battery technology, and AI, collaborating to develop a vehicle that is not only capable of vertical flight but also optimized for energy efficiency, safety, and ease of use. Central to AEROHT's EVTOL design is the integration of advanced AI systems that allow for autonomous navigation, route optimization, and collision avoidance, making the vehicle safe and reliable for both private and shared urban transit. By leveraging AI for real-time adjustments and decision-making, AEROHT's EVTOLs are designed to handle complex flight paths in dense areas, adapting quickly to changing conditions and ensuring a smooth, safe ride for passengers.

These EVTOLs are also intended to work in synergy with Xpeng's existing infrastructure. For example, they could be used in conjunction with ground-based electric vehicles, creating a seamless transition from land to air for users who need flexible, multimodal transport options. With the ability to operate independently of traditional road networks, EVTOLs developed by AEROHT could open up new corridors of transit that bypass typical bottlenecks, offering a faster, more direct path across cities. This capability to circumvent traffic congestion has the potential to reshape urban mobility, making commuting more efficient and reducing time spent on the road.

In addition to private passenger transport, AEROHT's EVTOLs could have applications in logistics, emergency response, and other specialized services that benefit from rapid, flexible deployment across urban and semi-urban areas. The versatility of EVTOLs makes them ideal for a range of functions, from urgent medical transportation to delivering goods in crowded city centers. This adaptability aligns with Xpeng's broader mission to develop intelligent, multipurpose vehicles that respond to the diverse needs of modern urban life.

Through AEROHT, Xpeng is positioning itself at the forefront of a rapidly evolving industry where air mobility is no longer science fiction but an emerging reality. As EVTOL technology matures, AEROHT's vehicles may become a practical option for daily commuting, offering a sustainable, efficient, and time-saving alternative to conventional modes of transportation. The work being done by AEROHT is a testament to Xpeng's commitment to pushing the boundaries of what's possible in AI-driven, electric mobility, paving the way for a future where people move freely between land and air as part of their everyday routine.

Xpeng's AEROHT division has taken a bold step toward making flying cars a practical reality with its hybrid flying car prototype, blending advanced aerodynamics with ground-based driving capabilities. This prototype is designed as a versatile vehicle that can transition seamlessly between driving on city streets and taking to the skies, offering a unique dual-functionality. Its hybrid nature combines the convenience of traditional car mobility with the agility and speed of aerial travel, making it suitable for short- to medium-distance urban and suburban commuting. The ability to switch between driving and flying modes allows this vehicle to circumvent road congestion and navigate directly over urban landscapes, redefining the concept of personal transportation.

Performance-wise, the hybrid flying car showcases impressive specs that make it well-suited for modern urban settings. The vehicle can carry up to six passengers and boasts a top flight speed of 224 mph (360 km/h) with a range of up to 311 miles (500 km) in flight mode, powered by a combination of electric motors for lift and hybrid engines for propulsion. This range allows users to cover significant distances without needing frequent recharging or refueling, making it a viable option for extended commutes and cross-city travel. On the ground, the car functions much like a traditional vehicle, enabling drivers to transition seamlessly from city streets to the skies when necessary. This flexibility offers a new level of mobility that adapts to the user's needs, whether they are navigating crowded urban roads or accessing quicker, open-air routes.

Xpeng has announced a projected release date for its hybrid flying car around 2026, with the vehicle expected to carry a price tag of approximately $279,000, making it one of the more accessible flying car options in development. To build excitement and prepare the public for this innovative mode of transport, Xpeng plans a series of demonstrations and trial flights, allowing consumers and city officials alike to experience the vehicle's potential firsthand. These demonstrations aim to showcase the car's performance, safety, and versatility, building trust and familiarity with the concept of flying cars as part of everyday life. By introducing the vehicle gradually, Xpeng intends to gather feedback and refine its design further, ensuring it meets both regulatory requirements and user expectations by the time it reaches market readiness.

Looking toward the future, flying cars like Xpeng's hybrid prototype could play a transformative role in urban transportation, providing solutions to some of the most pressing challenges in city mobility today. With urban populations continuing to grow, congestion and limited infrastructure have become significant obstacles, often leading to longer commute times and higher stress levels for city dwellers. Flying cars present a solution by opening up a new dimension of travel that operates independently of traditional road networks. In this vision, flying cars would create new aerial corridors above cities, bypassing bottlenecks and allowing people to move swiftly between destinations. This capability has the potential to reduce traffic, improve commute times, and make urban living more efficient and enjoyable.

In preparation for this future, Xpeng is planning pre-orders for the hybrid flying car as early as December. These pre-orders, along with planned demonstrations, mark the beginning of a phased introduction aimed at integrating flying cars into the urban landscape gradually. By allowing early adopters to secure a vehicle, Xpeng hopes to build a community of users who can provide insights into the practicalities of daily flying car use, from parking and charging to air traffic coordination and safety protocols. This early user feedback will be invaluable as Xpeng works with regulatory bodies to ensure safe and orderly implementation within city airspaces.

While flying cars may still seem futuristic, the technology and vision behind Xpeng's AEROHT division make it increasingly plausible for urban settings. In the coming years, flying cars like the hybrid prototype could become a regular part of urban transit systems, particularly for densely populated areas where road expansion is limited. The vehicle's hybrid design allows for seamless integration into existing transport networks, while its vertical takeoff and landing capabilities enable it to operate from rooftops, parking structures, and specially designated landing pads. These developments could shift urban planning, with cities potentially dedicating spaces for flying car operations, much like they did for ride-sharing and electric vehicle charging stations in the past.

By developing a hybrid flying car that combines the benefits of traditional driving with the flexibility of flight, Xpeng is setting the stage for a new era in urban mobility. The hybrid flying car prototype serves as a bridge between current ground transportation methods and the aerial mobility of the future, offering a glimpse into how people might move through cities in the decades to come. As Xpeng moves forward with this ambitious vision, the hybrid flying car could redefine personal transportation, making it faster, more adaptable, and more exciting than ever.

Chapter 8: Xpeng's Global Expansion and Market Strategy

Xpeng's expansion into international markets reflects its ambition to establish itself as a global leader in electric vehicles, AI technology, and sustainable transportation solutions. From its beginnings as a Chinese electric vehicle manufacturer, Xpeng has strategically entered key markets, building a strong presence in over 30 countries. This international growth is not only a testament to Xpeng's commitment to innovation but also a strategic move to meet the rising global demand for clean, efficient, and technologically advanced vehicles. As the global EV market continues to expand, Xpeng has positioned itself to compete with major automakers and emerging EV brands alike, drawing in customers with its unique combination of high-performance vehicles, cutting-edge AI integration, and commitment to sustainability.

Xpeng's expansion strategy is built on a carefully curated approach to market entry, focusing first on countries where demand for electric vehicles is already strong and infrastructure support for EVs is on the rise. By targeting regions with established or growing EV markets, such as Europe and parts of Asia, Xpeng can leverage its competitive advantages in AI-driven features, autonomous driving capabilities, and innovative design. In these markets, consumers are looking not just for any electric vehicle but for one that provides an elevated experience—one that blends performance with intelligence. This focus allows Xpeng to attract a segment of the market that values advanced technology and is ready to adopt the future of mobility.

Beyond merely selling cars, Xpeng is actively investing in the infrastructure needed to support its growing international customer base. This includes establishing service centers, charging stations, and customer support networks to ensure that new customers receive the same level of support and convenience as they would in Xpeng's home market. Currently, Xpeng operates over 145 after-sales service centers worldwide, a number that is expected to grow as the company expands. By ensuring that customers have access to maintenance, repairs, and charging infrastructure, Xpeng is building a foundation of trust and reliability, crucial for brand reputation as it enters new regions.

Xpeng's ambitious plans for international growth include expanding from its current presence in 30 countries to 60 countries over the next few years. This next phase of expansion aims to introduce Xpeng's products to new regions in North America, Europe, and additional parts of Asia and Africa, tapping into both established and emerging markets for electric vehicles. As part of this strategy, Xpeng is working closely with local governments and regulatory bodies to ensure compliance with diverse international standards, a key factor in building credibility and ease of access in different countries. This collaborative approach is designed to facilitate smoother entry into these new markets, allowing Xpeng to offer its full range of vehicles, services, and technologies on a global scale.

The drive to expand internationally is fueled by Xpeng's vision of becoming a household name in sustainable and AI-powered transportation worldwide. This vision aligns with the growing shift towards electrification across the automotive industry, as consumers and governments alike prioritize environmental sustainability and innovation. As more countries set ambitious targets for reducing carbon emissions and banning internal combustion engine vehicles, Xpeng is poised to become a primary choice for consumers seeking eco-friendly, tech-enhanced vehicles. This international footprint not only accelerates Xpeng's growth but also reinforces its role in advancing global clean energy initiatives.

By steadily increasing its reach, Xpeng demonstrates a commitment to bridging cultural and geographical boundaries through its technology. Its expansion represents more than just market entry; it is an effort to introduce a new level of mobility that emphasizes intelligent design, environmentally conscious engineering, and a seamless driving experience. As Xpeng pursues its goal to be present in 60 countries, it sets the stage for a future where advanced electric vehicles and intelligent transportation solutions are accessible to people all over the world. This commitment to a global vision underscores Xpeng's belief in a borderless approach to innovation, where mobility and technological advancement transcend regions, bringing sustainable progress to drivers and communities on every continent.

Xpeng's commitment to establishing a global service network is a crucial part of its strategy to provide high-quality, reliable support to customers around the world. With over 145 after-sales service centers already in operation, Xpeng has laid a strong foundation for ensuring that customers in international markets receive timely maintenance, repair services, and technical support. These service centers are strategically located to cover key regions in each market, ensuring that customers have convenient access to certified service professionals and parts, which is especially important for electric vehicles that may require specialized care. This infrastructure not only enhances the customer experience but also builds trust and brand loyalty as Xpeng introduces its vehicles to new markets.

As Xpeng continues its expansion into more regions, it plans to increase the number of these after-sales service centers significantly, reinforcing its commitment to customer satisfaction and support. By scaling its service network alongside market growth, Xpeng ensures that each new customer, whether in Asia, Europe, or North America, has access to comprehensive vehicle care. This commitment to service reflects a long-term strategy aimed at building lasting relationships with customers, which is essential in a competitive EV landscape. Additionally, Xpeng is exploring partnerships with local companies and service providers in various regions to enhance its support network further, adapting to local requirements and enhancing operational efficiency in each unique market.

Xpeng's aggressive expansion strategy and investment in global support infrastructure are positioning the company as a leader in the tech landscape beyond just electric vehicles. While Xpeng initially gained recognition as an EV manufacturer, its emphasis on AI-driven technology, autonomous driving capabilities, and sustainable mobility solutions have established it as a forward-thinking technology brand with influence that reaches beyond automotive circles. By pushing into diverse international markets with a sophisticated service network and technologically advanced vehicles, Xpeng is signaling its intent to lead the charge in the next generation of intelligent, eco-friendly mobility.

In a tech landscape dominated by Silicon Valley giants and traditional automakers, Xpeng's approach stands out for its ability to blend automotive engineering with cutting-edge artificial intelligence. Through innovations like its Turing AI chip, Kangai Level 4 autonomy, and EVTOL vehicles, Xpeng is contributing to advancements in multiple sectors, from robotics and AI to sustainable transportation. Its global expansion underscores this ambition, as Xpeng positions itself not just as an automaker but as a technology company dedicated to solving some of the most pressing mobility and environmental challenges of our time.

By establishing a robust service network and continuously expanding its reach, Xpeng strengthens its influence in the global tech arena, showcasing how intelligent and sustainable mobility solutions can be implemented on a worldwide scale. The company's presence in numerous countries gives it the opportunity to shape consumer perceptions of electric and autonomous vehicles, setting standards for what future mobility can look like in terms of both performance and convenience. Through its strategic infrastructure and focus on innovative technology, Xpeng is not only advancing electric vehicle adoption but also paving the way for a broader transformation in global transportation and urban living.

As Xpeng continues to grow, its presence in the global tech landscape is likely to inspire other companies to pursue similar advancements, particularly in regions where the adoption of sustainable technologies is still emerging. By establishing itself as a reliable, innovative brand with a strong support network, Xpeng is redefining what it means to be a global technology leader, creating a lasting impact that reaches far beyond its origin as an EV manufacturer.

Chapter 9: China's Role in Xpeng's Rapid Growth

Government support has been a crucial factor in propelling Xpeng's advancements in AI and robotics, aligning with national and regional goals to position China as a global leader in high-tech innovation. China's government has heavily invested in the fields of AI and robotics, recognizing these technologies as foundational to future economic growth, industrial efficiency, and international competitiveness. This support comes in the form of funding, subsidies, infrastructure development, and favorable regulatory policies, all of which create an environment where companies like Xpeng can thrive. These investments aim to accelerate the development of AI and automation across industries, with Xpeng at the forefront of integrating these technologies into practical, real-world applications.

One of the key ways the government fosters growth in AI and robotics is through substantial research and development grants, which support companies in exploring advanced technologies that require intensive capital investment. For Xpeng, these grants provide the financial resources necessary to push the boundaries of autonomous driving, robotics, and EV innovation. By accessing government funding, Xpeng can accelerate the development of technologies like the Turing AI chip, Kangai Level 4 autonomous driving platform, and its humanoid robot, Iron, all of which contribute to the company's ambitious vision for a tech-driven future. This funding is especially important for the development of proprietary technologies, which require long-term investment and extensive testing to meet the high standards of safety and performance required in consumer products.

In addition to direct funding, the government has also invested in creating specialized industrial zones, tech parks, and infrastructure that support the growth of AI and robotics companies. These innovation hubs serve as collaborative spaces where tech firms, research institutions, and academic organizations can work together on projects, share resources, and accelerate development. For Xpeng, access to these hubs provides opportunities to collaborate with top researchers and industry experts, fostering an environment where the latest advancements in AI and robotics can be integrated into the company's products. This access to a talent-rich ecosystem has proven invaluable in advancing the company's capabilities in areas like autonomous driving, machine learning, and robotic intelligence.

Regulatory support is another critical component of the government's strategy to boost AI and robotics. China's government has created policies that encourage rapid deployment and testing of new technologies, which is particularly important for companies developing autonomous vehicles and robotics, where stringent regulations can otherwise slow down progress. By establishing flexible regulatory frameworks that allow for pilot testing and phased implementations, the government enables Xpeng to bring its innovative solutions to market more quickly. This regulatory flexibility allows Xpeng to conduct real-world testing of its autonomous driving technology, humanoid robots, and EVTOL vehicles, all under government-approved protocols that ensure safety while fostering rapid innovation.

Furthermore, the government's emphasis on advancing AI and robotics aligns with China's broader goals to reduce reliance on foreign technology and cultivate homegrown solutions that can compete on a global scale. This focus on technological self-reliance encourages companies like Xpeng to prioritize proprietary technology, such as the Turing AI chip, which is custom-designed to support Xpeng's specific needs in autonomous vehicles and robotics. This strategic focus not only boosts Xpeng's competitiveness in international markets but also contributes to the national goal of positioning China as a technology leader in emerging industries. By developing its own advanced technology, Xpeng helps to reduce dependency on international suppliers, aligning with national objectives and enhancing its ability to scale globally.

The influence of government support is evident in the speed and scale at which Xpeng has been able to innovate and expand. From its AI-driven robot taxis and EVTOL ambitions to the advanced robotics embodied by Iron, Xpeng's ability to bring these futuristic projects closer to reality is rooted in an ecosystem where government and industry work hand-in-hand. This partnership accelerates innovation and reinforces the company's commitment to integrating AI and robotics into everyday life, setting an example for other companies in China and beyond.

Overall, government investment in AI and robotics has been instrumental in empowering Xpeng to pursue groundbreaking projects that might otherwise be prohibitively expensive or difficult to realize. By providing financial resources, fostering collaboration, and creating an adaptable regulatory environment, the government has laid a foundation for Xpeng and similar companies to become leaders in the global tech landscape. This synergy between government policy and corporate ambition not only drives Xpeng's success but also advances China's standing as a powerhouse of technological innovation, poised to shape the future of AI, robotics, and sustainable mobility on the world stage.

China's robust manufacturing and innovation infrastructure has played a pivotal role in enabling Xpeng's ambitious projects, providing a unique ecosystem that supports rapid development, efficient production, and large-scale deployment of advanced technologies. The country's manufacturing capabilities are among the most sophisticated in the world, with specialized zones dedicated to high-tech industries, extensive supply chains, and a skilled workforce that can adapt to the complexities of producing next-generation products. This well-developed ecosystem gives Xpeng a significant advantage, as it can leverage local resources, materials, and expertise to bring its electric vehicles, AI-powered robots, and EVTOLs from concept to reality at a competitive pace and cost.

China's manufacturing network is designed to support both mass production and customization, allowing Xpeng to meet market demands for quality and innovation. From electric vehicles to autonomous driving systems and AI chips, Xpeng benefits from close access to suppliers who specialize in everything from semiconductors to electric batteries, minimizing delays and streamlining production. Furthermore, China's focus on smart manufacturing and automation aligns closely with Xpeng's objectives. This alignment makes it easier for Xpeng to incorporate its AI-driven processes within local factories, utilizing a blend of human expertise and robotic assistance to maintain quality control and efficiency. By tapping into this ecosystem, Xpeng can rapidly prototype, test, and iterate its products, ensuring that each innovation meets the company's high standards before going to market.

China's strategy to position itself as a global leader in technology and AI has created an environment where companies like Xpeng can thrive and compete on an international scale. Recognizing the potential of AI and robotics to reshape industries, China has invested heavily in developing technologies that can challenge those from other tech powerhouses like the United States, South Korea, and Japan. This national drive is part of a broader vision to secure China's place as a key player in the global tech landscape, promoting self-reliance, advancing local capabilities, and fostering companies that can make an impact worldwide. Xpeng plays a crucial role in this movement, standing at the intersection of AI, robotics, and sustainable mobility—a combination that aligns perfectly with China's technological ambitions.

Xpeng's innovative projects, such as the Turing AI chip, Kangai Level 4 autonomy, and humanoid robot Iron, are prime examples of how Chinese companies are advancing technology on multiple fronts. Xpeng's ability to integrate these advanced features into practical applications showcases the potential of China's tech sector to produce world-class, cutting-edge solutions. Moreover, Xpeng's EVTOL (electric vertical takeoff and landing) vehicles, developed through its AEROHT division, highlight China's ambition to lead in urban air mobility and expand into new areas of transportation. These projects not only enhance Xpeng's profile but also contribute to China's reputation as a center for high-tech innovation that can compete directly with the best the world has to offer.

China's strategic support for high-tech sectors, including AI, robotics, and autonomous vehicles, is a key factor in Xpeng's success. Through subsidies, research grants, and government-backed initiatives, China creates an environment where companies are encouraged to pursue ambitious, high-risk projects that push the boundaries of current technology. This support has allowed Xpeng to experiment and innovate in areas like AI-powered autonomous systems and customized chipsets for robotics, which are essential for realizing its vision of a fully integrated, AI-driven future. Xpeng, in turn, has embraced this opportunity, channeling the support into creating technologies that are not only competitive but also unique to China's tech landscape.

As China continues its push to be a leader in AI and advanced manufacturing, Xpeng serves as a key example of how Chinese companies can drive global innovation. With its sophisticated manufacturing ecosystem, government backing, and a growing domestic talent pool, China is positioning itself as a formidable competitor on the global tech stage. Companies like Xpeng play a dual role in this strategy—fostering national pride in China's technological achievements and establishing China's reputation as a source of high-quality, technologically advanced products on the world market.

In summary, China's strong manufacturing and innovation infrastructure has been instrumental in enabling Xpeng to pursue its ambitious projects with confidence. The country's strategy to advance AI, robotics, and tech places Xpeng at the heart of a movement aimed at reshaping the global tech landscape. By supporting companies like Xpeng, China not only accelerates its path to technological leadership but also demonstrates how integrated ecosystems of manufacturing, innovation, and policy can drive meaningful advancements in AI-driven mobility and sustainable technology worldwide.

Chapter 10: The Future of AI-Driven Innovation at Xpeng

Looking beyond 2024, Xpeng's long-term vision is centered around pioneering advancements in AI, robotics, autonomous mobility, and sustainable transportation. Xpeng has established itself as a leader in innovation with its remarkable achievements in electric vehicles, autonomous driving, and robotics. However, its ambitions extend far beyond the current landscape, as the company envisions a future where smart, AI-driven technologies seamlessly integrate into every aspect of daily life, transforming urban spaces, redefining transportation, and enhancing the human experience.

Future AI Day events will likely showcase Xpeng's continuous push to innovate, with a focus on breakthrough technologies that align with its goal of creating a cohesive ecosystem of intelligent solutions. As the industry evolves, these events will provide Xpeng with a platform to unveil the next generation of products and technologies that leverage AI to empower sustainable and efficient urban living. These AI Day showcases will emphasize Xpeng's commitment to transparent, cutting-edge innovation, demonstrating not only advancements in autonomous driving but also the integration of AI across multiple domains, from robotics and chip technology to smart urban mobility solutions like EVTOL vehicles.

In the realm of autonomous driving, Xpeng's long-term vision involves reaching and refining Level 5 autonomy, where vehicles can operate entirely without human intervention under all conditions. Building on the success of its Kangai Level 4 autonomy, Xpeng aims to expand its capabilities to deliver a fully autonomous experience that transforms both personal and shared transportation. By advancing to Level 5, Xpeng envisions a future where robotaxis and autonomous vehicles are integral to urban transit systems, reducing traffic, enhancing safety, and improving accessibility for people of all mobility levels. In the coming years, Xpeng may reveal advancements in AI-driven sensor technology, vehicle-to-vehicle communication, and AI algorithms that enhance predictive decision-making, all contributing to safer, more reliable self-driving technology.

Xpeng's vision for AI-powered robotics also extends into areas that go beyond manufacturing and customer service. Robots like Iron have already shown the potential to support complex tasks, but the next phase could involve AI-driven robots entering specialized fields such as healthcare, logistics, and hospitality, where intelligent, adaptable machines could perform valuable tasks with high levels of precision and care. With continued development in the Turing AI chip, Xpeng is likely to enhance Iron's capabilities, making it more versatile, responsive, and capable of adapting to an even broader range of environments and roles. As AI and machine learning evolve,

Xpeng's robots could become more autonomous, capable of sophisticated interactions and personalized responses that bring value in both professional and personal spaces.

In urban mobility, Xpeng's AEROHT division reflects the company's ambition to lead the future of air transportation. The development of EVTOL vehicles is part of a broader strategy to create a new dimension of urban travel, reducing road congestion and offering faster, more flexible ways to move through cities. Beyond 2024, Xpeng may explore further integration of air mobility solutions with ground-based transportation, creating a network of multi-modal transit options that provide seamless transitions between land and air. As EVTOL technology advances, Xpeng's flying vehicles could be equipped with AI-driven navigation and route optimization systems that allow for autonomous or semi-autonomous flights, making urban air travel safe, efficient, and accessible. Future AI Days may reveal enhancements in EVTOL energy efficiency, autonomous flight control, and urban infrastructure support, moving closer to realizing flying cars as a staple of city life.

Additionally, Xpeng's long-term vision includes sustainable innovation, ensuring that each technological advancement aligns with environmental goals. With China and many other nations setting aggressive targets for carbon neutrality, Xpeng is poised to lead in creating clean energy solutions that reduce emissions without compromising performance or user experience. This focus on sustainability will likely drive Xpeng to explore further improvements in battery technology, energy efficiency, and the use of renewable resources within its products and manufacturing processes. Future AI Days may highlight advancements in sustainable materials, innovative battery recycling methods, and new energy solutions that support both ground and aerial vehicles, reinforcing Xpeng's commitment to eco-friendly innovation.

In the coming years, Xpeng's AI Day events will serve as a glimpse into a future where AI-driven technology, robotics, and sustainable mobility become integral to urban life. These events will showcase not only the technological prowess of Xpeng but also its dedication to creating solutions that enhance quality of life, support environmental goals, and make intelligent mobility accessible on a global scale. As Xpeng continues to expand its reach and refine its technologies, it will be well-positioned to lead a new era in intelligent transportation, where the lines between automotive, robotics, and AI seamlessly blend into a cohesive and transformative vision for the future. Through continuous innovation, Xpeng aims to redefine what's possible, shaping a world where advanced technology supports, connects, and empowers people in meaningful, sustainable ways.

Xpeng is on an ambitious path toward achieving full autonomy in its vehicles, with plans that reach far beyond the automotive industry. By perfecting autonomous driving, Xpeng aims to revolutionize how people and goods move within cities, reshaping urban landscapes and paving the way for transformative applications across various industries. As autonomous driving technology progresses from Level 4 to the ultimate goal of Level 5 autonomy, Xpeng envisions a future where its vehicles operate independently in any environment, eliminating the need for human intervention. This shift will not only redefine personal transportation but also open doors to new industries that rely on safe, reliable, and intelligent mobility solutions.

Xpeng's journey to full autonomy involves refining every aspect of its technology, from advanced sensor systems and AI algorithms to vehicle-to-infrastructure connectivity. Achieving Level 5 autonomy requires vehicles to perceive and respond to complex, dynamic environments seamlessly. Xpeng's investment in custom-designed AI chips and neural networks, like the Turing AI chip, reflects its commitment to building the processing power and adaptability needed for true autonomy. By integrating high-definition mapping, predictive modeling, and real-time data analysis, Xpeng is developing vehicles that can handle unpredictable elements, such as construction zones or sudden weather changes, with the same reliability as a skilled human driver. This level of autonomy will enable applications beyond private transportation, including robotaxis, autonomous logistics, and even emergency response vehicles, all of which can contribute to smoother, safer urban operations.

In the context of urban life, Xpeng's advancements in autonomous driving promise to reshape city landscapes by reducing congestion and streamlining transportation systems. Autonomous vehicles (AVs) can be part of a coordinated network that moves people and goods efficiently, minimizing traffic jams and lowering emissions by optimizing routes and reducing idle time. The rise of robotaxis, powered by Xpeng's Kangai Level 4 and future Level 5 autonomous platforms, could offer an accessible and convenient alternative to personal car ownership, especially in densely populated areas. By providing on-demand, autonomous transport options, Xpeng can contribute to a significant reduction in the number of cars on the road, allowing cities to reclaim space currently dedicated to parking and improve air quality through reduced vehicle emissions.

Xpeng's innovations are also set to transform logistics and delivery services, where autonomous driving could enhance efficiency and reduce costs. Self-driving delivery vehicles could operate 24/7, ensuring that goods move quickly from warehouses to retail locations or directly to consumers without delays. The potential for automated logistics extends to industries that require consistent, round-the-clock supply chains, such as healthcare and manufacturing. By reducing dependence on human-driven vehicles, autonomous logistics networks can address labor shortages, increase efficiency, and optimize supply chains in ways that make urban environments more resilient and adaptable to change.

Beyond the road, Xpeng's focus on robotics and AI expands the impact of its technology into areas that directly shape urban living and everyday interactions. Humanoid robots like Iron, equipped with Xpeng's proprietary AI, could play a vital role in customer service, retail, and healthcare, where personalized assistance and adaptability are essential. In retail, Iron could assist customers with product information and demonstrations, creating a shopping experience that is more interactive and informative. In healthcare, AI-powered robots could assist in non-critical tasks, such as delivering supplies, guiding visitors, and supporting staff with repetitive duties. This integration of robotics into everyday environments could streamline operations, improve service quality, and allow human workers to focus on more specialized and complex tasks.

Globally, Xpeng's vision for full autonomy and intelligent robotics has the potential to set new standards for urban mobility and technology-driven lifestyles. As autonomous and AI-driven solutions become more widespread, Xpeng's innovations could serve as a model for other cities and companies looking to integrate these technologies. Autonomous vehicles and robots could work in tandem to create smart cities where transportation, delivery, and services operate cohesively, reducing friction points and enhancing quality of life. Xpeng's EVTOL (electric vertical takeoff and landing) vehicles add another dimension to this vision, offering a new mode of aerial transport that bypasses traditional road networks and opens up alternative routes within urban spaces. EVTOL technology could be used for short commutes, emergency services, or fast logistics, making cities more versatile and adaptive to high-density populations.

In the long run, Xpeng's commitment to full autonomy and AI-powered solutions could drive a global shift toward sustainable, intelligent cities where technology supports human needs on multiple levels. By embedding smart, autonomous systems into urban infrastructure, Xpeng's work may lead to a future where mobility, customer service, and logistics are seamlessly integrated, making urban life more efficient, convenient, and environmentally conscious. As these technologies gain traction on a global scale, they will likely influence city planning, infrastructure investment, and workforce development, setting a standard for how technology can elevate urban living.

Ultimately, Xpeng's innovations are more than just advancements in transportation—they represent a shift in how society interacts with technology in everyday life. Through its pursuit of full autonomy, robotics, and sustainable mobility solutions, Xpeng is positioning itself as a pioneer of the future, showing how AI and autonomous systems can enrich urban living, improve environmental outcomes, and redefine our approach to mobility on a global scale. This forward-thinking vision highlights the broader impact of Xpeng's work, suggesting a future where cities are smarter, transportation is safer, and technology serves as a reliable partner in building a connected, sustainable world.

Conclusion

Xpeng's transformation from a dedicated electric vehicle manufacturer to a leader in AI-driven technology marks a remarkable journey fueled by bold vision, relentless innovation, and a commitment to reshaping the future of mobility and urban life. Beginning with the goal of producing efficient, intelligent EVs, Xpeng quickly expanded its scope to encompass a much broader vision, leveraging artificial intelligence, robotics, and advanced autonomy to explore the boundaries of what technology can achieve. From its Turing AI chip and the Kangai Level 4 autonomous driving platform to humanoid robots like Iron and the AEROHT division's EVTOL flying vehicles, Xpeng has evolved into a company that consistently pushes the frontier of what's possible. This transformation reflects not only the company's ambitious goals but also its capacity to drive change in the global tech landscape.

The impact of Xpeng's innovations reaches far beyond the automotive industry. With its humanoid robots, autonomous vehicles, and EVTOL technology, Xpeng is poised to redefine daily life and set new standards for convenience, efficiency, and sustainability. The potential applications of these technologies—autonomous urban transport, AI-powered customer service, and versatile flying vehicles—signal a future where intelligent machines support people in new, transformative ways. Xpeng's work could soon lead to cities with reduced congestion, optimized logistics, cleaner air, and smarter, more adaptive infrastructure. The integration of AI-driven robots into everyday settings, the rise of fully autonomous robotaxis, and the introduction of flying vehicles all speak to a future where advanced technology becomes seamlessly embedded into the fabric of urban life.

As Xpeng continues its journey, the world is watching, eager to see how its ambitious vision unfolds. For readers intrigued by Xpeng's progress, staying informed about its advancements promises a front-row seat to the next wave of technological breakthroughs. Each AI Day event, new product announcement, and milestone achieved will not only highlight Xpeng's growth but also showcase innovations that could shape the lives of people worldwide. In an era where technology is evolving at an unprecedented pace, Xpeng's story is a compelling reminder of the possibilities that lie ahead when creativity and technology come together to tackle real-world challenges.

The journey is far from over. Xpeng's vision for the future—one filled with intelligent, interconnected, and sustainable solutions—holds promise for cities, industries, and individuals alike. As Xpeng continues to redefine the limits of mobility, artificial intelligence, and robotics, we can expect it to lead the way toward a future that's smarter, cleaner, and more connected. So, to those who see the potential of AI to elevate our lives and the role of companies like Xpeng in realizing that potential, now is the time to follow along and witness a transformation that could shape the world for generations to come.

www.ingramcontent.com/pod-product-compliance
Lightning Source LLC
Chambersburg PA
CBHW070422240526
45472CB00020B/1150